CW01333651

Southern EMU Pictorial

Images from the Transport Treasury archive

Hector Maxwell

TOTEM PUBLISHING

© Images and Design: The Transport Treasury or as credited 2025. Text: Hector Maxwell.

ISBN 978-1-913893-57-6

First published in 2025 by Transport Treasury Publishing Ltd. 16 Highworth Close, High Wycombe, HP13 7PJ
Totem Publishing, an imprint of Transport Treasury Publishing.

The copyright holders hereby give notice that all rights to this work are reserved.
Aside from brief passages for the purpose of review, no part of this work may be reproduced,
copied by electronic or other means, or otherwise stored in any information storage and
retrieval system without written permission from the Publisher. This includes the illustrations
herein which shall remain the copyright of the copyright holder.

www.ttpublishing.co.uk

Printed in Tarxien, Malta by Gutenberg Press Ltd.

Front cover: Former 3-SUB No 4309 seen at Victoria. This unit entered service with the Southern Railway as No 1310 in August/September 1925 and was subsequently provided with an additional coach, c1945-6, giving it the new designation 4-Sub. The 'P' headcode indicates a Victoria - Beckenham Junction via Streatham Hill service. ('P' was also used for services to Crystal Palace High level from Blackfriars.) Set No 4309 was withdrawn on 7 May 1960. *(Kichenside 1921 / Transport Treasury)*

Frontispiece: 4-SUB No 4343 on the Western Section, again augmented from a former 3-car set in 1945-6, seen passing Raynes Park with a Shepperton - Waterloo service, via Earlsfield. The augmenting vehicle is clearly identified by the differing width profile. This unit remained in service until 22 April 1961. *(Kichenside 1940 / Transport Treasury)*

Rear cover: 'Modern traction' in the shape of 2-EPB No 5747 also at Victoria. Some of the type, Nos 5651-5684, were built on recovered 2-NOL underframes, others including that depicted had standard BR Mk1 underframes. The first BR 2-EPB saw the light of day on 31 December 1953 and entered service in 1954 initially on the South London line and on Wimbledon - West Croydon services. In their final days they were designated Class 416. *(Kichenside 1957 / Transport Treasury)*

Introduction

Pick up almost any edition of the regular Ian Allan 'ABC' railway 'spotters' books from the 1960s and one name will almost certainly appear as the credit against an image of an electric unit or a carriage - 'G J Kichenside'.

Geoffrey Kichenside was born in 1932 in Kenton, North West London. His interest in railways originated from his maternal grandfather who was on the Southern Railway. Even at the tender age of seven his grandfather explained a lot about signalling – what the signals were and where they were positioned.

After leaving The Lower School of John Lyon, which was related to Harrow School, he was employed by Ian Allan compiling the 'Locospotters' publications. He then progressed to magazine editor of 'Model Railway Constructor', 'Railway World' and eventually 'Modern Railways'.

Then in the early 1970s Geoffrey received 'an offer he could not refuse' and moved to the West Country to take charge of the railway list at David & Charles, Newton Abbot: readers of a certain age may well remember the D&C LNER coaches parked at the west end of Newton Abbot station painted in their distinctive brown and sunshine yellow shade.

He had also been involved in passenger-carrying miniature railways as part of the Ian Allan Group, and the move to Devon provided similar opportunities in a tourist business, as a result miniature railways became a major part of his activities, including the creation of the 7¼" Gorse Blossom Miniature Railway Park set in 10 acres of land at Bicklington (Devon) close to the A38 which he ran with his wife Paula for 18 years. He was also active in a full-size railway as part of the Dart Valley Group.

Geoffrey's input to and as a writer of railway books was considerable, and this excludes whilst acting as editor / compiler of various Ian Allan ABC and annuals. In addition to his individual output, Geoffrey also collaborated with Alan Williams in books on signalling; together the pair made for a formidable combination.

Geoffrey's negatives passed to Transport Treasury in 2024, unfortunately without much if anything in the way of annotation.

It is hoped that what follows will be of interest to enthusiasts of Southern EMUs, who might even recall the bouncy ride on sprung cushions, the occasional feeling of claustrophobia in a compartment full of bodies, or the gentle sway of a 'Cor' unit on the curves of the Portsmouth direct.

Geoffrey did not appear to have much if any interest in the modern EMU scene, and to be honest, who could blame him.

He lives in retirement with his wife in South Devon.

Here we have a suburban 2-Nol set No 1868 at Twickenham and dating from February 1936. The set was originally designated 'third' but changed to 'second' from 3 June 1956. The 'Nol' classification referred to the train as having 'No Lavatory'. A total of 76 two-car sets units entered service (Nos 1813-1890) between 1934 and 1936. As was Southern Railway, and indeed Southern Region policy for many years, they were an amalgam of the old and the new. Underframes and electrical equipment was new whilst the bodies had started off as LSWR steam hauled stock. The motor third brake vehicles had originated as 48' eight-compartment thirds, one compartment was removed and replaced by a half-compartment for the guard, the remainder devoted to luggage space and the driver's cab; the alterations and additions show up well in the photograph. The second vehicle in the set this time was a composite again rebuilt, with the necessary addition of a driving cab. The '8' headcode is unofficial and likely to be unofficial or even graffiti. ('8' implies a Waterloo - Portsmouth Harbour service not calling at Havant although there may be some doubt as to this.) Unit No 1868 remained in service until 30 November 1957. *(Kichenside 1859 / Transport Treasury)*

Another set that started off as a 3-Sub* and is seen here with the addition of an additional (Bulleid) trailer is No 4331 - now designated 4-Sub. At peak times two 3-Sub units would have had a non powered 2-coach set sandwiched between them through wired to allow one motorman to control the whole train. This worked well excepting the fact it required shunting of the two-car non-powered units usually at the end of the morning peak and before the commencement of the evening peak. Having a fully formed train was seen as a better option hence the augmentation previously referred to. The set is seen stabled at Hampton Court with the 'H; indicating a service to and from Waterloo. On the Eastern section service, the letter 'H' referred to services from London Bridge to Victoria via Tulse Hill and Streatham Hill, OR London and Herne Hill, Orpington to Holborn Viaduct or Sevenoaks to Holborn Viaduct. (*Kichenside 1644 / Transport Treasury*) (Note although the term '3-Sub' is used throughout this book, it is more accurately an unofficial - enthusiasts term.)

Augmented 4-Sub this time No 4511. This set was one of 18 units reformed and renumbered in the 1950s, usually due to the original units being involved in accidents or with vehicles withdrawn due for the need of heavy repairs. According to Geoffrey's (brief) notes, the vehicle seen had started life as an LBSCR composite. The headcode 'H' was described in the previous view. Alongside is a 'modern' 4-Sub No 4295. The location of this view and that opposite is Hampton Court. *(Kichenside 1662 / Transport Treasury)*

No 4557 displays its origins well - ex LSWR compartments. Throughout the life of these trains oil tail lamps were a daily feature, the smell of paraffin something that has been missing from the railway for some decades now. Considering the intensive electric suburban service operated by the SR and the short dwell times at stations, it is remarkable to consider how luggage might have been carried and indeed the time available to load and unload items. Notice also the guard's side lookouts. *(Kichenside 1666 / Transport Treasury)*

A pair of Sub units at Hampton Court, one at least identified as No 4528 - note too the difference in driver's cab handrails - ex LSWR stock on the right. Coach S8739S, a motor brake third, is a former LBSCR vehicle modified and incorporated into what was then a unit in the series 1757 - 1768. This set entered service in early 1930 and had a life a few months short of 30 years being withdrawn on 12 December 1959. *(Kichenside 1674 / Transport Treasury)*

The unit number of this set is likely 1748, again consisting former LBSCR vehicles. We can at least positively identify the motor brake second as S8717S. Note especially the window ventilators with the added new front portion of the vehicle beyond the furthermost passenger compartment, giving the impression the bodywork may have dropped slightly. *(Kichenside 1678 / Transport Treasury)*

No 4301, with motor brake trailer No S8130S at the front. This is another augmented set - the third vehicle being to a wider body profile - from a 3-Sub to a 4-Sub. At the rear is the modern version of a suburban set - all wider stock - to the same Sub classification. Notice too that periscope look outs are provided above the guard's compartment compared to a side duckett. This is one of several images taken at Hampton Court. *(Kichenside 1800 / Transport Treasury)*

The steel sheeting forming the portion of this motor brake second shows up well in this view of set No 4331 at Hampton Court. This was one of the early electric sets dating from 1925/6, augmented to four cars in September 1945 and in service until 20 January 1962. As was also typical for electric units of the time, a windscreen wiper is only fitted to the driver's window, the nearside window hinged so that it would open allowing the driver to reach out and place the appropriate (route) stencil in the centre panel. Note too the whistle on the offside. *(Kichenside 1798 / Transport Treasury)*

Unit 4134, dating from LSWR days leaving Richmond. Component parts / vehicles in this unit dated back as far as June 1898 and it appears some vehicles were in service as late as 12 December 1959. As with most Southern EMUs, reforming took place on a number of occasions and in consequence the records can become confusing. The late Ted Crawforth for example, who was a former SR driver based at Waterloo for many years, took great pains to record as much detail on EMU formations from the offices at Croydon whenever the opportunity presented itself. His notes show that vehicles within this set were substituted in 1952, 1956 and again in 1958, before what was left of unit No 4134 ceased to exist at the end of the following year. *(Kichenside 1902 / Transport Treasury)*

This view at Clapham Junction shows up the 'bullet' type front of the original LSW units well. The straight sides also contrast somewhat with the Bulleid brake third / second coach in the background. Unit 4531 had started life as LSWR set No E24, then SR No 1224 and when augmented as No 4531 as seen here. In its original three-car formation were LSWR vehicles Motor Brake Trailer 6736, Composite 7574, and Motor Brake Composite 7212; these numbers retrospectively changed to SR Nos 8036, 9383, and 8762. Under British Railways the 'S' prefix and suffix were added. This set survived until 10 November 1956. The 'H' stencil was referred to on page 5.
(Kichenside 1878 / Transport Treasury)

Another LSWR unit, this time, we think No 4170. There is also another comparison with the Bulleid 9-compartment trailer alongside. The Bulleid suburban vehicle was built with some compartments to first-class width although not operated as composites (except one in set 701). Coach No S8035S would indicate this was originally LSWR set No E23, SR No 1223 and latterly BR No 4170.
(Kichenside 1880 / Transport Treasury)

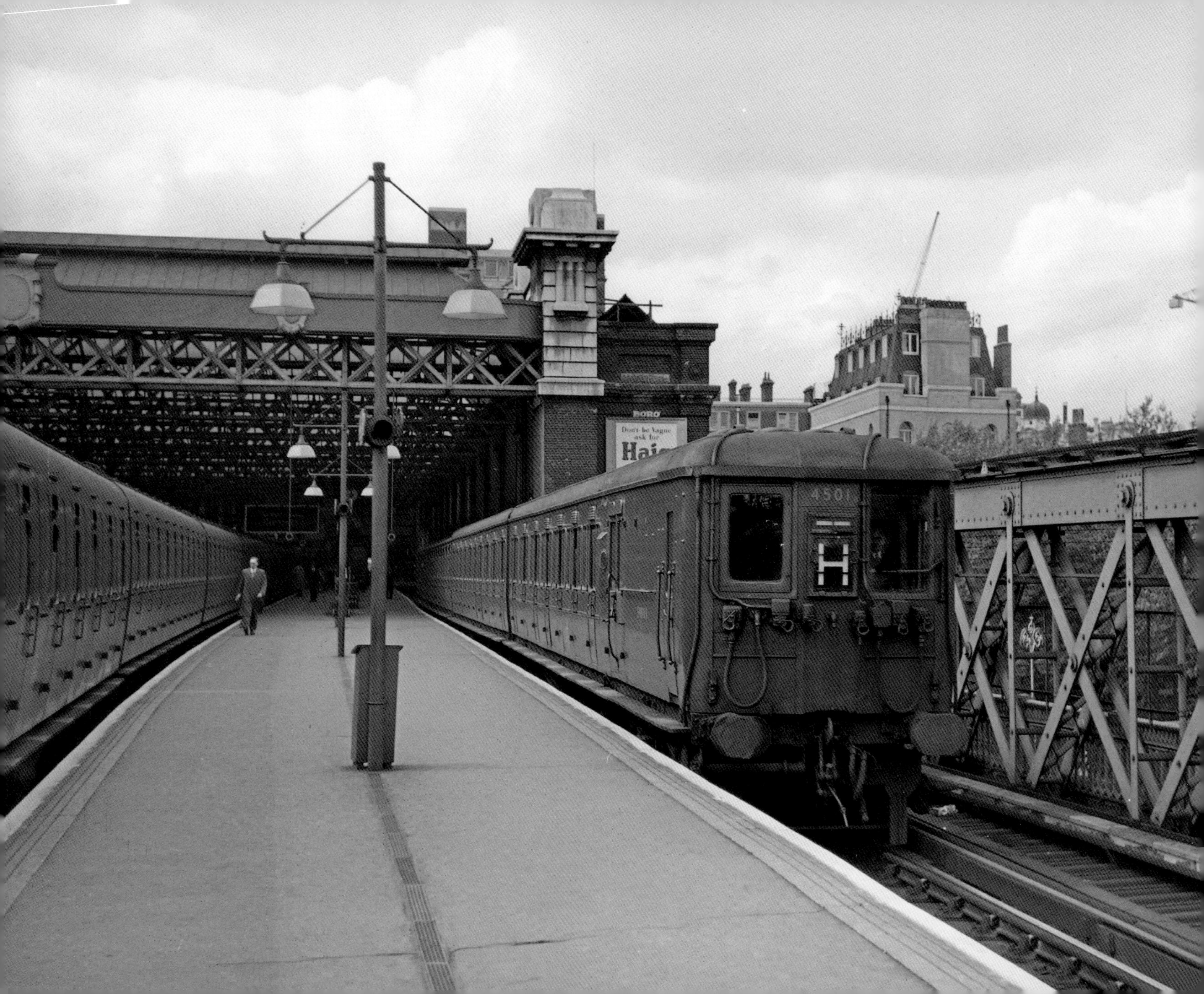
BORO
Don't be Vague
ask for
4501
H

Opposite: Charing Cross has unit 4501 awaiting departure. The 'H' headcode and tilde (bar above the letter) indicates a Charing Cross to Hayes via the Ladywell loop train. The later numerical equivalent was '34'. Alongside is a modern Sub or Epb unit. *(Kichenside 1883 / Transport Treasury)*

Above: No 4176 at the distinctive curve of Clapham Junction. Notice the bracket into which a destination board might be fitted on the side of the guard's compartment. An inward opening door to the driver's compartment, common to all units, allowed for ease of access from ground level although a duplicate door handle which could be reached from ground level might have been a worthy addition. *(Kichenside 1885 / Transport Treasury)*

Suburban operations at Waterloo. Two Sub units, one ex LBSCR and one LSWR That nearest the camera is No 4532, formerly SR No 1731 which would be in service until the end of 1959. According to the Kichenside notes the second unit is No 4501. Alongside is the modern equivalent No 4366. The '30' is for a Waterloo - Hampton Court train. *(Kichenside 1863 / Transport Treasury)*

2-Nol No 1873 at Twickenham. As an example of the changes made over the years, of the 76 sets of this type, in 1944 four trains had been modified with the removal of the coupé compartment; by 1946 this number had risen to 33. This unit had entered service in 1936 and was withdrawn on 6 July 1957. Intended for main line stopping services, they were renowned for their deep comfortable seating and over the years could be found throughout most of the Central and Western section electrified network. *(Kichenside 1863 / Transport Treasury)*

A further view of 2-Nol No 1873 again at Twickenham and positively sparkling in the sunshine and berthed in the bay platform used for rugby traffic for the nearby stadium. The front end paintwork has faded somewhat with the information on the data plate hardly visible. With no tail lamp or route number / letter shown, we may reasonably conclude the set was parked on layover.
(Kichenside 1852 / Transport Treasury)

No 4142 sporting the original bullet end design, perhaps looks somewhat dated against the modern background of Twickenham. Running as a 4-Sub, it may be seen that all four vehicles now match in profile achieved by inserting a centre car from perhaps a similar withdrawn unit. This set survived until 16 June 1956. *(Kichenside 1854 / Transport Treasury)*

Comparison of front ends at Hampton Court. Driving brake trailer No S9815S gives us the clue that this is 4-Sub No 4573 (February 1949 to February 1956) or 4501 (July 1956 to January 1960) depending on the date of the photo (which we do not have). Very few 3-car units survived into BR days and in consequence it is doubtful any carried the lion over wheel emblem and the carriage numbers with the suffix S as well as the prefix S. The suffixes only appeared c1950. The unit is attached to a modern, and has to be said austere Sub unit. Careful examination of the couplings showed that both trains were indeed coupled and so proves the point that although externally visually dissimilar, the various Sub unit were comparable and did indeed operate conjoined in the one train.
(Kichenside 1850 / Transport Treasury)

LSWR set No 4215 in the sidings at Hampton Court and hopefully awaiting its next duty. We say 'hopefully' as the view was clearly taken in the 1950s but also the time when these early units were being withdrawn - this set saw the end on 8 July 1955. Note the set number painted on the solebar; a sensible step for a driver / guard to identify his train at the start of duty but it seems to have been applied only to the former LSWR trains - was it in fact used on all units and has simply not been noticed in the other images?
(Kichenside 1856 / Transport Treasury)

4-Sub unit No 4324 in the same location, Hampton Court, as seen previously in the image on page 11 but with a totally different augmentation trailer. Construction wise these trailer vehicles were built in similar fashion to the 'Sheba' 4-sub units, Nos 4101-4110 being a wood framed body and canvas roof and with one fewer compartment than the cramped 'Sheba' units. The inward opening driver's door is also seen to advantage. Buffing and coupling gear at the end of the sets was standard and comparable with other stock although only a single centre buffer was used between cars making up the sets. Note also the 'Ladies only' compartment immediately next to the guard. Although compartments for Ladies had been present on some lines as far back as the 1840s, calls for more came about after the conviction of an army Colonel for indecent assault on a 21-year-old woman on a service from Portsmouth to Waterloo in June 1875. The victim forced to flee through the compartment's only door, which opened externally, leaving her balancing on the running board outside the moving train and clutching the door handle until the train came to a halt. There were also converse opinions such as in November 1936 when Transport Minister Leslie Hore-Belisha* was asked in the Commons about them. 'I am informed that there are carriages marked 'Ladies only', but that the ladies prefer not to use them, he told the House. Even so some persisted almost to the 1960s although by now nearly all restricted to trains operating on the suburban systems. With the introduction of open carriages on suburban services, separation was no long possible. *(Kichenside 1850 / Transport Treasury)*

* As an aside, Leslie Hore-Belisha gave his name to the lights at the side of the road identifying a pedestrian crossing, the 'Belisha Beacon'.

Buffer stop view of 4-Sub No 4302. A tail lamp has yet to be added and it may be noted the driver has left his newspaper in the cab! Again too a variation in handrails. *(Kichenside 1812 / Transport Treasury)*

Above and opposite: What appear to be four 2-car sets on layover at Twickenham. Aside from No 1890, also the highest number allocated to a 2-Nol unit, no details are given of the other sets. Mention has been made of the irony of having an oil tail lamp on an electric train and yet should the power fail the oil lamp would remain lit. The position of the actual lamp brackets, one of either side on the lower portion of end, enabled this essential component to be added from track level. *(Kichenside 1822 (left) and 1832 (right) Transport Treasury)*

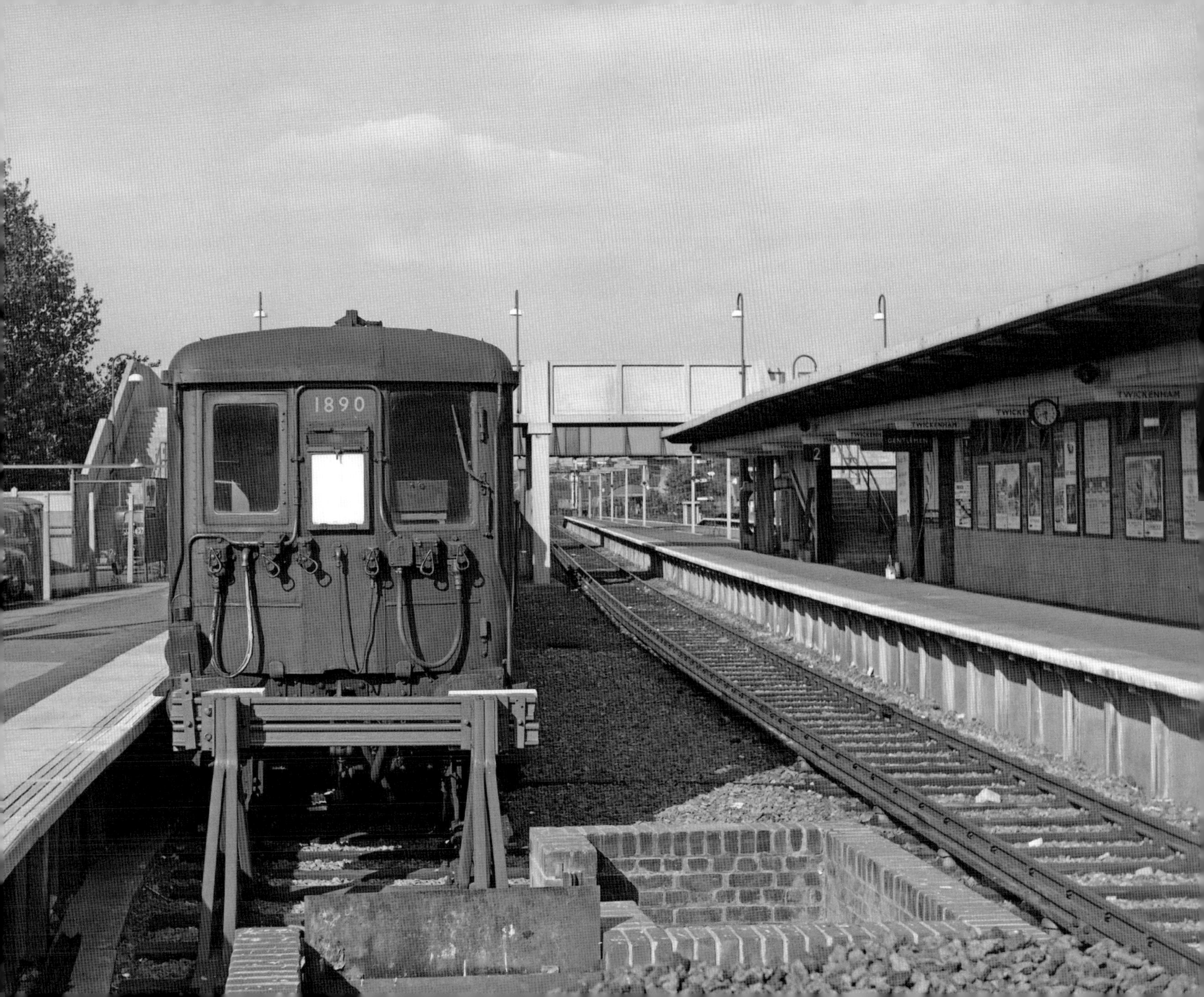
1890
TWICKENHAM
TWICKENHAM
GENTLEMEN
2

Both pages: Individual carriage detail but without information either from GK or from the individual vehicles - almost. What we can say with certainty is that the then 'standard' Bulleid vehicle is No S10183S which dates from October 1948. It is clearly being used as an augmentation vehicle but without detail as regards which set; although we can say this one is of all-steel construction. We also have exterior compartment detail of a pre-group conversion, the Smoking / No Smoking signs contradictory.
(Kichenside 1837 (left) and 1672 (right) Transport Treasury)

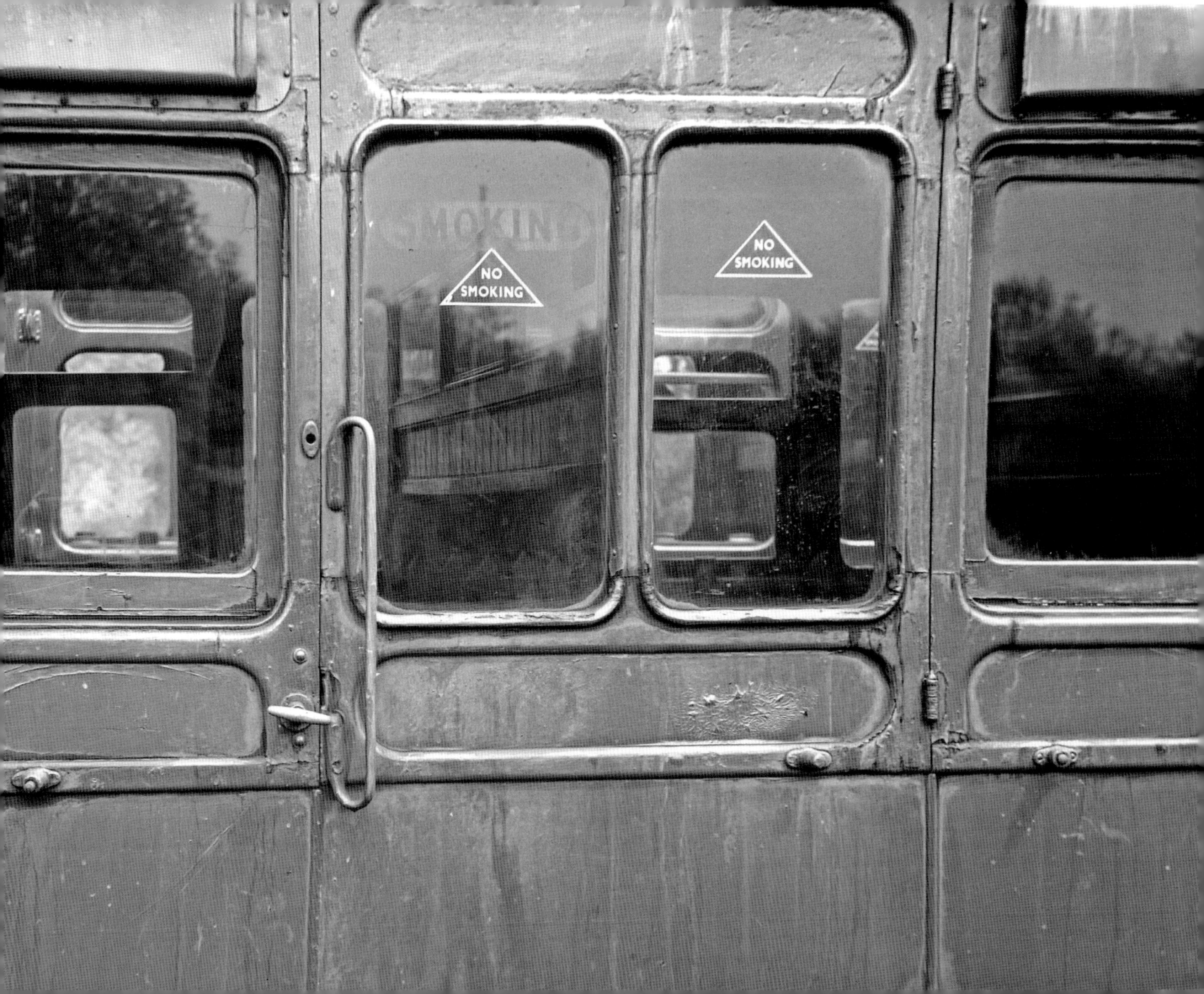
NO SMOKING
NO SMOKING

Top: LBSC S9475S is a former LBSCR vehicle that Ted Crawforth's records show was included in the formation of Set No 4557. A 10-compartment vehicle it could accommodate either 80 persons as four-a-side seating or 100 if five-a side, any standing passengers are not counted. Records shown in the excellent David Brown book on 'Southern Electric' have this coach as having first been converted for multiple unit operation sometime between February and April 1930. Notice the patch repairs on several doors and the complete absence of a builders plate on the solebar. (*Kichenside 1644 / Transport Treasury*)

Bottom: S8746S from unit No 4564 is a former LBSCR vehicle converted to a motor brake third. Consistency in SR days meant regardless of the origins of the vehicle to be adapted there would still be eight compartments of third class accommodation plus a brake compartment. Again a modern intermediate vehicle is attached. (*Kichenside 1670 / Transport Treasury*)

Opposite: A conversion from LBSCR stock, S8739S a motor brake third attached to a second unit which has its origins on the LSWR; either No 4528 or 4555. Again there are items previously mentioned, the contradictory Smoking / No Smoking signs, Ladies compartment, drivers cab handrails and guard's lookouts - side or roof. The difference in roof heights may also be noted. The vehicles shown on both pages are at Hampton Court. (*Kichenside 1676 / Transport Treasury*)

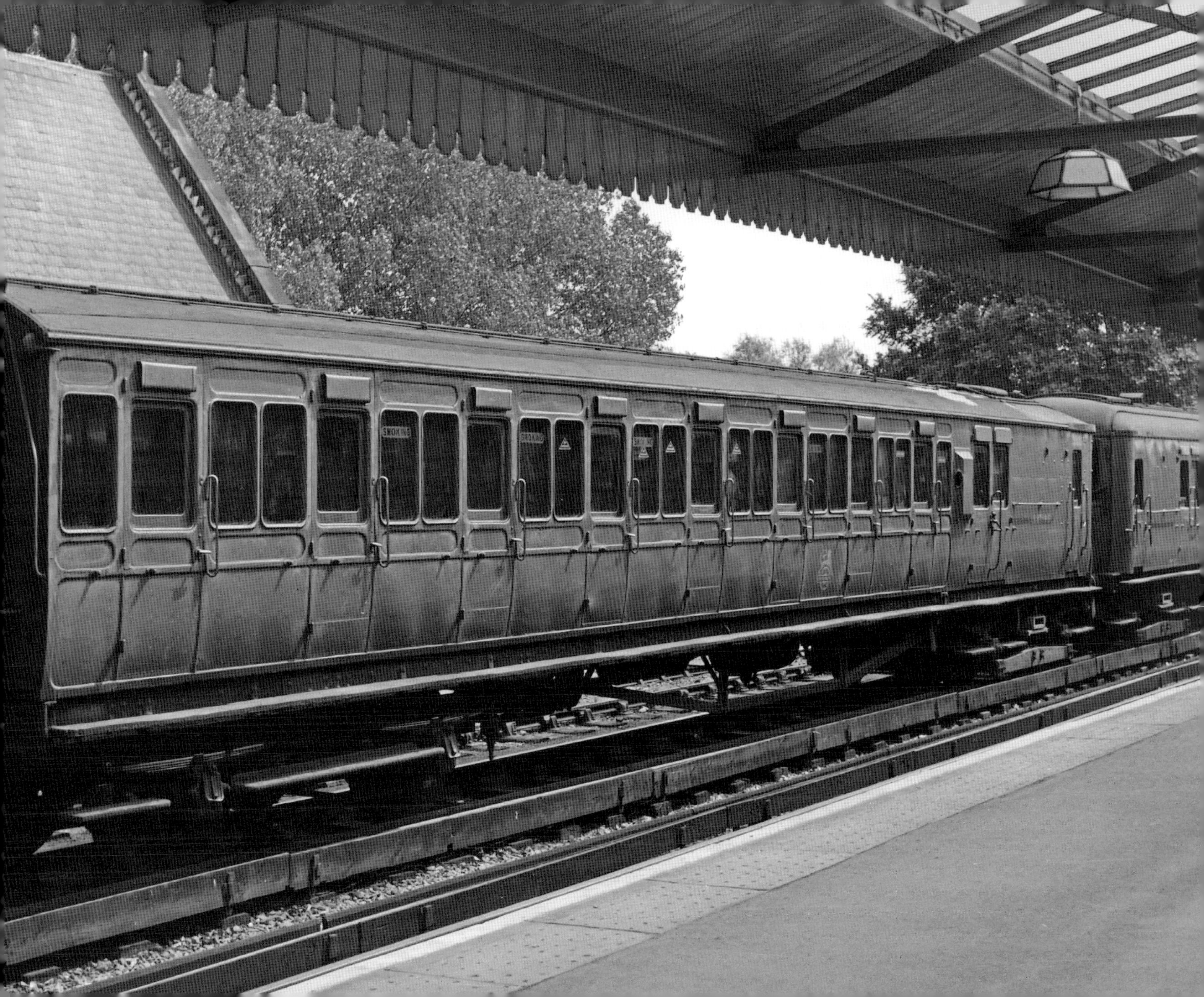
SMOKING

S9458S was built for EMU working in 1925 and is in set No 4325. The slab sides are similar to some former SER stock and as with all Sub stock of the period, non-corridor. Again there is evidence on some patch repairs to the lower portions of the doors; possibly water ingress from the droplights causing rot. It is interesting to pause a moment to view the poses of the passengers, one might even sum it up as 'resigned indignation': an acceptance that was just part of the daily ritual. The lady is in her own world whilst the gentlemen with the newspaper is nowadays an endangered species, broadsheets having given way to tablets and smart-phones. There is a data plate on the end indicating a tare weight of 27 tons. (*Kichenside 1728 / Transport Treasury*)

Sister vehicle S9459S from set No 4324 in the sidings at Hampton Court. Most of the same comments from the previous page apply here except to also say this vehicle was first placed into unit 1309 when first converted. It would survive on paper at least until 18 March 1961, but as with all official data, it may not have been in service, perhaps stored, for a period before this. Worth mentioning is that the circular BR insignia as applied to EMUs did not appear on the intermediate vehicles. (*Kichenside 1732 / Transport Treasury*)

Two for the price of one, Unit 4314 head on and intermediate coach No S9604S from No 4331. No 4314 had started life as set No 1299 whilst the coach opposite is a downgraded composite first seen in set No 1501. One aspect noticeable throughout the images in this book is the lack of external graffiti as was commonplace in later years whenever units were stabled out of use even for short periods. (*Kichenside 1743 / Transport Treasury*)

Driving trailer of set 1879 - S9936S. The lack of a brake compartment meant an additional compartment was possible. Notwithstanding the headcode having no relevance, the stencil '18' may well have indicated a recent attachment as per, 'Staines - Weybridge (detached from or attached to service via Richmond)', 'Waterloo - Chessington South', or 'Staines - Windsor (detached from or attached to service via Richmond'. Readers may notice the white dot at the ends of both sets. This indicated that brake modifications had been made to the unit as a result of the 1952 Guildford accident where a train ran away. Unmodified units were not allowed to work solo. The white spot indicated that a safety cut out had been added and was applied to 2-Bil and 2-Nol units. There is no mention of a similar modification to 2-Hal units. (*Kichenside 1826 / Transport Treasury*)

Top: S9782S a former LSWR brake composite modified here as all third LSW motor brake third. Note the early 'cycling lion' BR emblem referred to previously. (*Kichenside 1876 / Transport Treasury*)

Bottom: Close up of the ownership insignia and the windows the recalled red triangle notice. Livery throughout was green and in BR days standard red with white flecks was the standard moquette used for seating without centre armrests, (*Kichenside 1869 / Transport Treasury*)

Opposite: A good image showing the differing width profiles. As was ever the case passenger preference was mixed, some preferring the perhaps more traditional older units and others the austere look of the modern. (*Kichenside 1867 / Transport Treasury*)

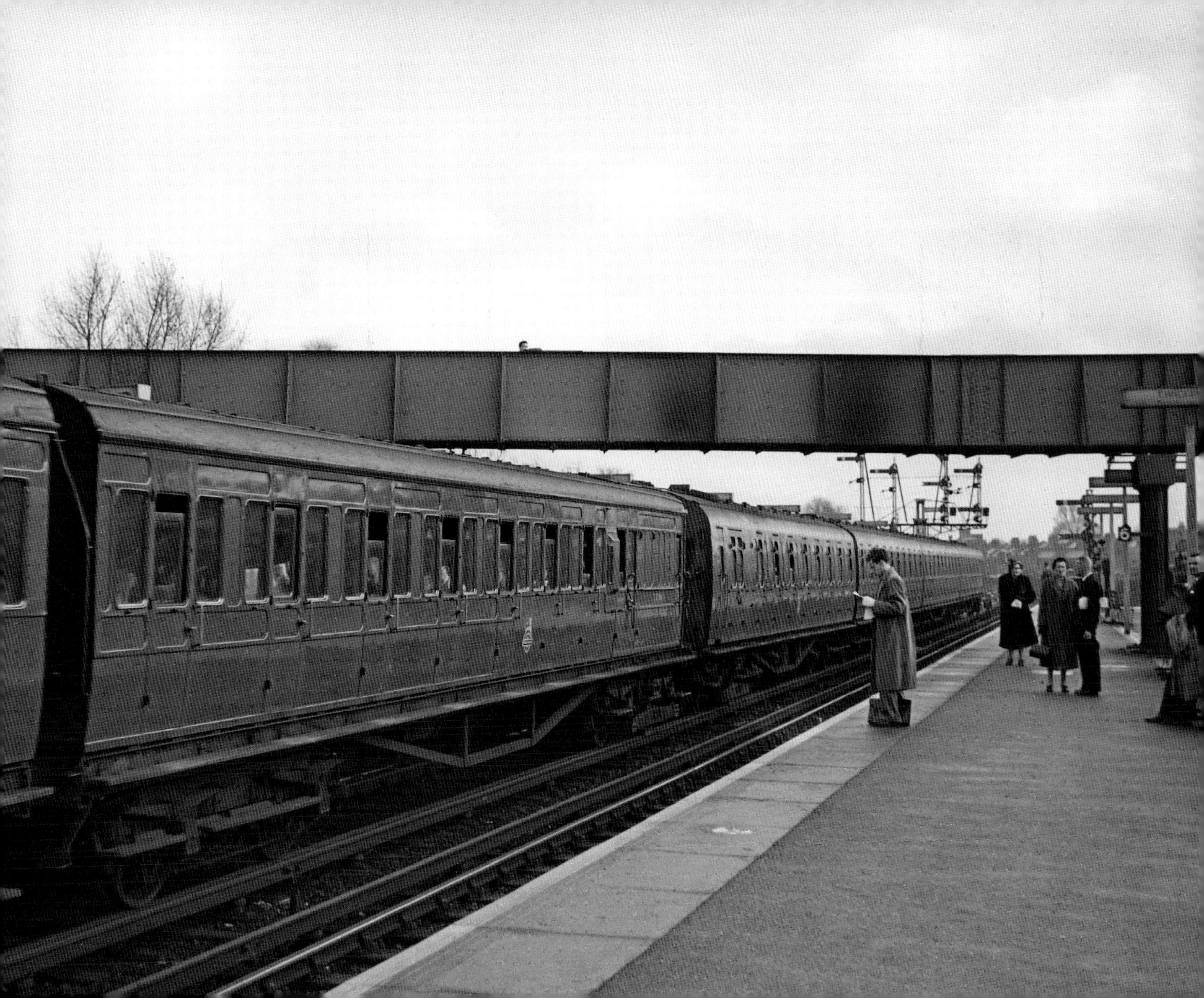

S9657S converted for EMU operation in July 1929 and withdrawn from service on 7 November 1959. There had been progressive withdrawals of the pre-war subs the last of these going by the end of 1962. (*Kichenside 1847 / Transport Treasury*)

Stabled units / stock the location unfortunately not recorded by GK. Identifiable is S9656S so this may well be unit No 4254. This is believed to have originally been as part of unit 1780 - although different sources also refer to set 1730. Of note and worth comparing with other images where they might be seen, are the axlebox covers from the three pre-group companies. (*Kichenside 1843 / Transport Treasury*)

Driving trailer, no guard's compartment, from a 2-Nol set and with its first class compartment clearly marked. The location is believed to be the west end of Surbiton. (*Kichenside 1834 / Transport Treasury*)

Another ex LSWR vehicle, again compare the axlebox covers. This is S9333S at Clapham Junction seen here as part of a 4-Sub all of similar vehicles. The bogie stepboards hark back to much earlier days but were a feature of the design and thus retained, although it must be said apart from those at the driving cab ends, they serve little or no useful purpose. (*Kichenside 1838 / Transport Treasury*)

Two seemingly identical units, No S9914S part of 2-Nol No 1884. Notice the bogie step boards to discourage using part of the collector gear as a foot step. (*Kichenside 1829 / Transport Treasury*)

2-Nol No 1869 having recently arrived at Richmond - notice the ex LNWR Ocrlikon set in the background on a North London line service. Signalmen part of whose responsibility was to observe trains as they passed, would have noted the door handle of the second compartment is not fully horizontal although the actual door appears fully closed. (*Kichenside 1836 / Transport Treasury*)

A pair of 1925 built SR units showing the connections between. It was of course necessary for the shunter to go in between the two units to attach the coupling and air lines ('palm' type connectors) but the electrical connection for MU working could be made at platform level if necessary and from either side. The blank droplight on the offside of No S8131S, 4Sub No 4302, is a puzzle and would have been expected to have been clear glass. (*Kichenside 1654 / Transport Treasury*)

Crossing the River Thames at Richmond. (*Kichenside 1632 / Transport Treasury*)

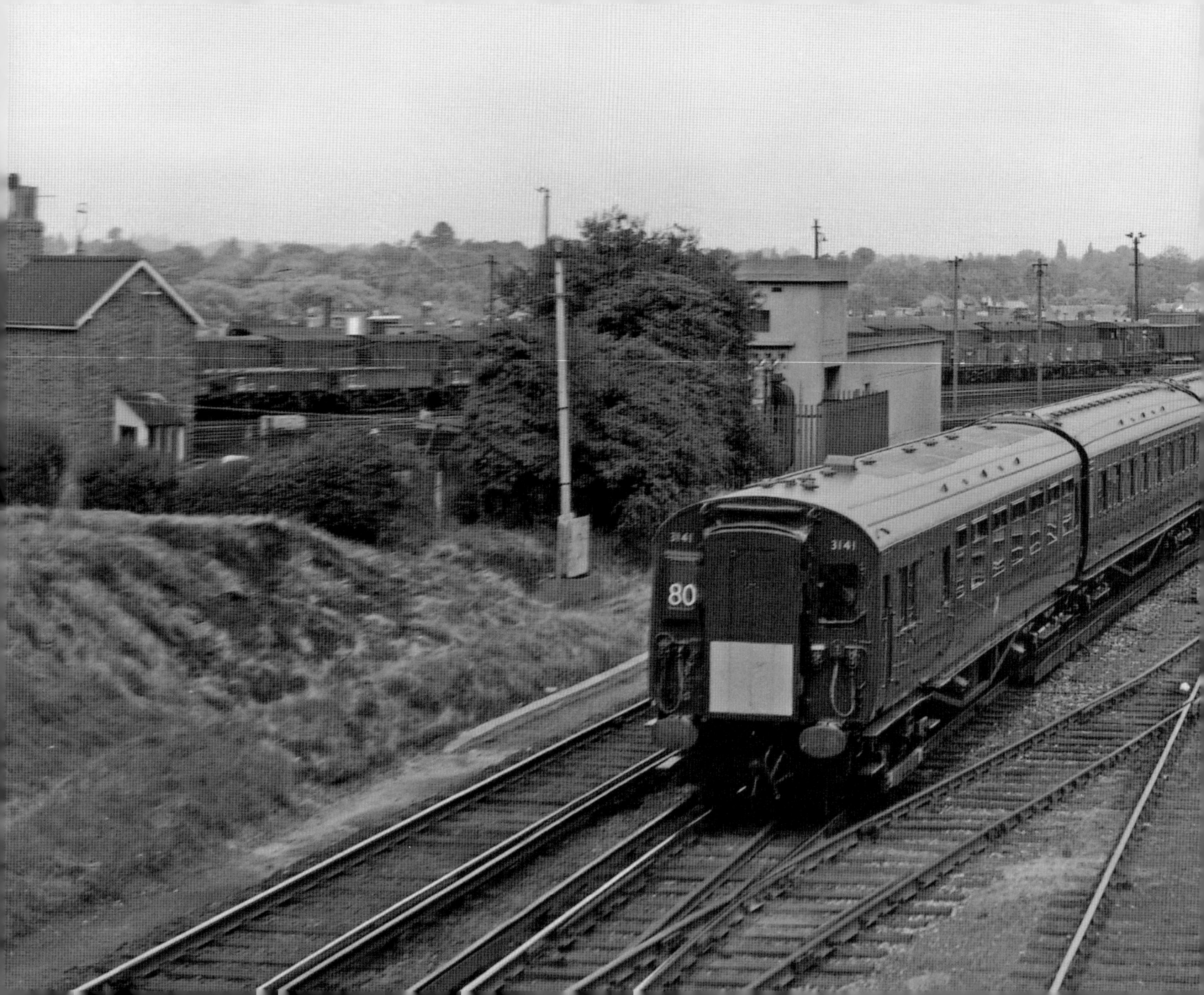
3141
80

Moving away from units intended for suburban and stopping services for a moment, we come to perhaps one of the best known of the electric types, the 4-Cor sets (and their variants 'Res', 'Buf' 'Gri' etc). This is 4-Cor No 3104, part of a set of three units, the middle one a 'Gri' (Griddle) unit making up '80' which is a Waterloo - Portsmouth Harbour service seen here taking the line south towards Guildford at Woking. *(Kichenside 1917 / Transport Treasury)*

Another of the main line sets of the period were the twenty 6-Pul sets, originally numbered 2001-2020 but renumbered 3001-3020 at the beginning of 1937. The formation of these trains was a mixture of first and third class plus a Pullman kitchen composite. At the same time three additional units, 2041-2043 (later 3041-3043) were built the difference being these were primarily all first class and catered for the up and down business services to and from Brighton and London Bridge. Post WW2 they were downgraded to the same duties as the majority 6-Pul sets although with a lower seating capacity. This is set 3011, which incorporates Pullman *Naomi*, and is seen at Haywards Heath. The numeric code '52' refers to a Victoria - Ore service via the Quarry line and Eastbourne, (post 1967 and long after these trains had been withdrawn it was re designated as Victoria - Newhaven Harbour / Seaford again via the Quarry line. *(Kichenside 1630 / Transport Treasury)*

Composite S11790, part of set 3012 dating from December 1932 (previously 2012) with Pullman *Bertha* behind passing Haywards Heath. No S11790S has the yellow stripe at cantrail level to indicate first class. On the motor coach a pick up beam will be noted on the rear bogie. *(Kichenside 1639 / Transport Treasury)*

Above: The end of the train seen on the previous page with a composite next to the Pullman and then third class accommodation beyond. Composite No S11789S may be identified. (*Kichenside 1774 / Transport Treasury*)

Right: The hustle and bustle of Clapham Junction is exemplified here with 4-Sub No 4573 entering with a 'Kingston Loop'; Waterloo to Waterloo via Earlsfield, Teddington and Richmond service. Unit 4573 dated from February 1929 and it is believed lasted until 18 February 1956 - again alternative data gives 1955 and 1960. (*Kichenside 1668 / Transport Treasury*)

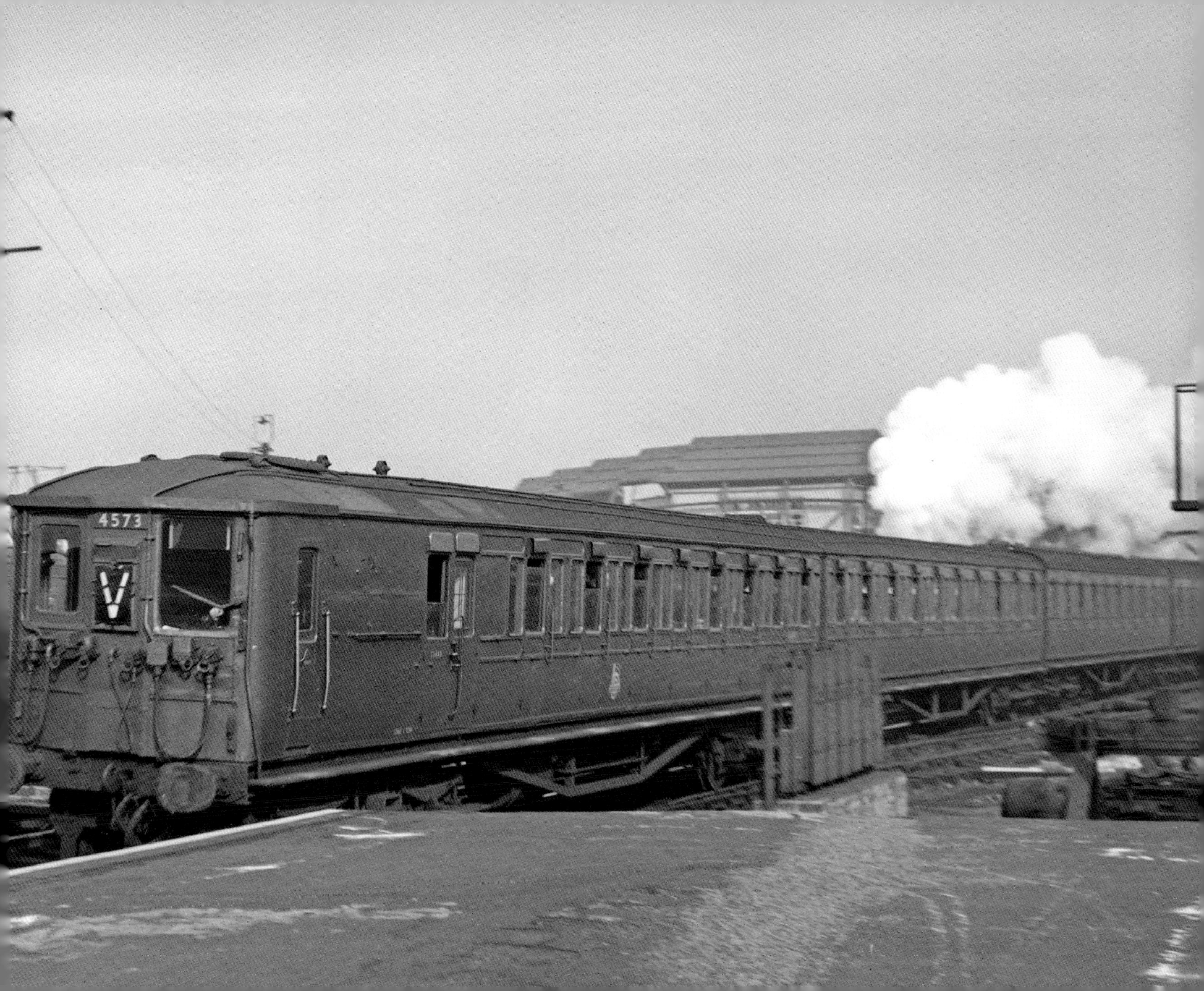
4573
V

This page, the comings and goings at Raynes Park.

Top: A pair of 2-Bil two-car units; 'Bil' meaning 'bi-lavatory'. These sets were of all new construction. Although there are alternative meanings, the '32' designation at Raynes Park indicates a service from Alton. *(Kichenside 1634 / Transport Treasury)*

Bottom: Eight car suburban service probably preparing to stop at the staggered platforms of Raynes Park. The rear set is No 4352, an augmented 3-Sub. *(Kichenside 1934 / Transport Treasury)*

Opposite - Down Central Section working at Haywards Heath. Set 3036, is a 6-Pan. Initially catering was provided including the services of a Pullman steward, this was undertaken in an SR vehicle with a kitchenette, counter and restricted range of light refreshments. Post-WW2, the pantries fell out of use except on one or two peak-hour trains and the units in effect became 6-Cor albeit not classified as such. After a life 29 years, these sets were taken out of service from 1964 onwards. *(Kichenside 1646 / Transport Treasury)*

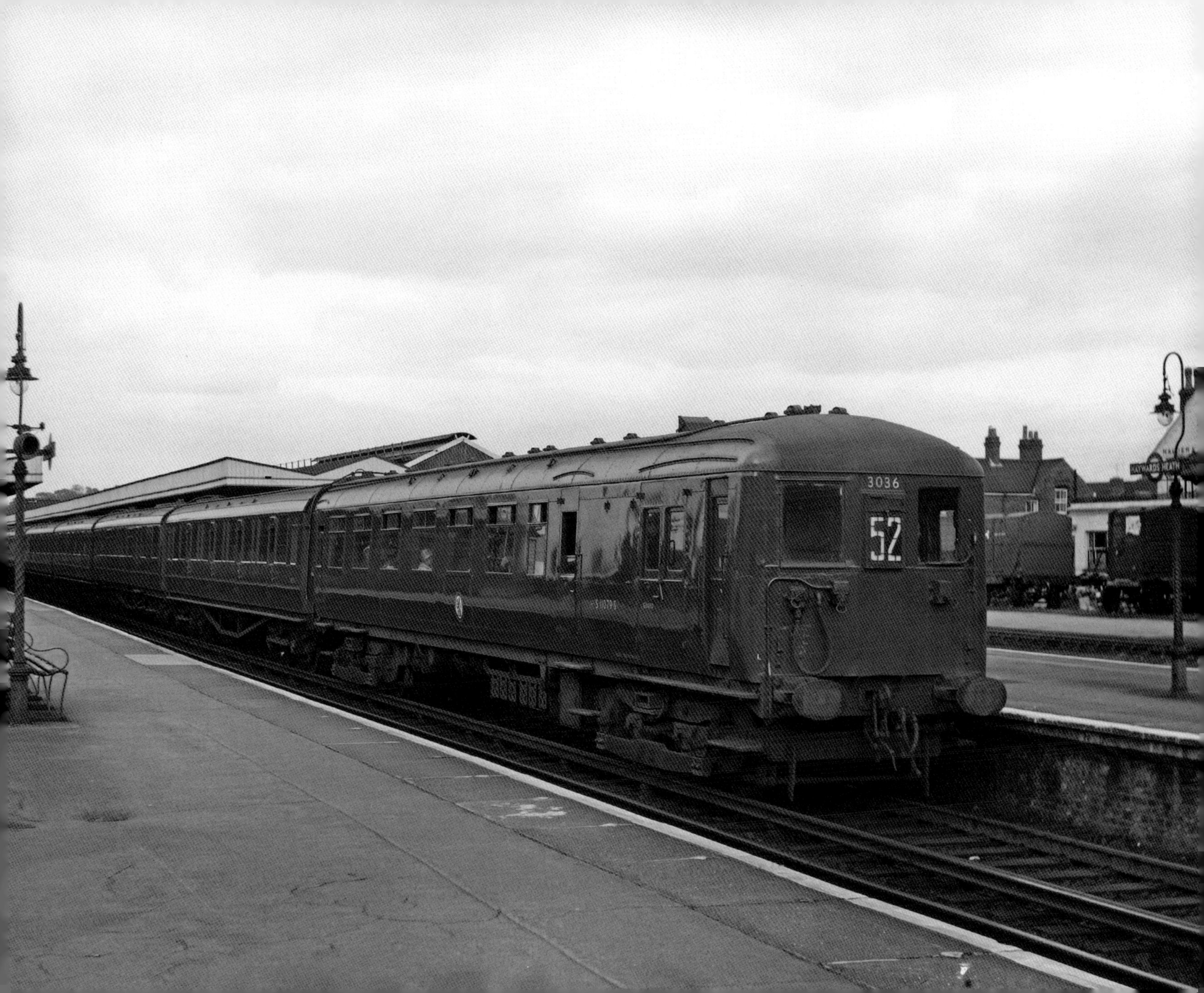
3036
52
HAYWARDS HEATH

4-Lav set No 2923 (built as No 1923 but renumbered in 1937) arriving at Three Bridges with a London Bridge to Brighton slow service. Built for Brighton line services, the two centre vehicles were composites, one with a lavatory - hence the type designation. Most of these units had a service life in excess of thirty years with slight changes in formations made in consequence of accident damage. Unit 2923 was built in June 1932 and was one of the last survivors remaining in use - on paper at least - until 5 April 1969. *(Kichenside 1658 / Transport Treasury)*

A type of electric unit we have not encountered previously is the 2-Hal type, illustrated here with unit 2664. A total of 76 of these two-car trains were built in the first half of 1939 for the Gillingham and Maidstone electrification. A further 16 sets were produced at the end of 1939 although the exact reason why they were ordered is unknown, however, they were used to bolster the Reading/Guildford via Ascot trains in WW2 due to the heavy military traffic to Aldershot. According to the excellent SEMG website, the luggage compartments might carry perishables such as milk churns and fish, the latter likely to leave a reminder of their presence for a time afterwards! Generally between 1955 and 1959 (but see image on the next page), it is thought one first-class compartment was downgraded on the sets allocated at the time to the Eastern Section but was reinstated as first class when the sets moved to the Central and Western Sections. Where a first was rendered third, the designation was removed form the doors and the armrests sewn into the seat-backs. This now provided for four-a-side seating with those in the know quickly occupying these compartments as they afforded greater legroom. *(Kichenside 1692 / Transport Treasury)*

Seen at Havant, the characteristic oval windows indicate a Hal, this is unit No 2650. Of note is the recessed driver's door which in turn displays the body width to maximum advantage. Despite displaying a small yellow panel at the front which indicates an early 1960s view, excepting with those sets working on the Eastern section, the first class accommodation were retained until withdrawal. This unit lasted until April 1970. *(Kichenside 1646 / Transport Treasury)*

One of the ever popular 2-Bil sets, represented by No 2035 at Havant. These trains had a composite first and a third trailer, a side corridor in both afforded access to the compartments. A lavatory was provided in each coach. These sets could be seen on most of the electrified Western Section lines; initially on Portsmouth line services, then a further batch for the Mid Sussex routes, and a final batch for the Reading lines and the associated Ascot to Guildford route. Service '60' was a semi-fast service in either direction between Brighton and Portsmouth. Notice on this set the patch repairs to the roof. One set, No 2090 was retained for official preservation with all the others out of service by July 1971. *(Kichenside 1704 / Transport Treasury)*

This time we see a pair of 2-Bils leaving Havant with a short Portsmouth - Chichester working. All four lines through the station are electrified, yet out of sight to the left 80- year old steam engines in the form of the 'Terrier' tanks were working the Hayling Island services until 1963. *(Kichenside 1910 / Transport Treasury)*

Woking Junction with a Waterloo - Portsmouth & Southsea stopping service and obviously routed via Worplesdon rather than the 'new line' to Guildford. This high angle image gives a good view of the guard's periscopes, roof ventilators and other equipment. Notice too the rain strips above the driver's and guard's access doors. *(Kichenside 1908 / Transport Treasury)*

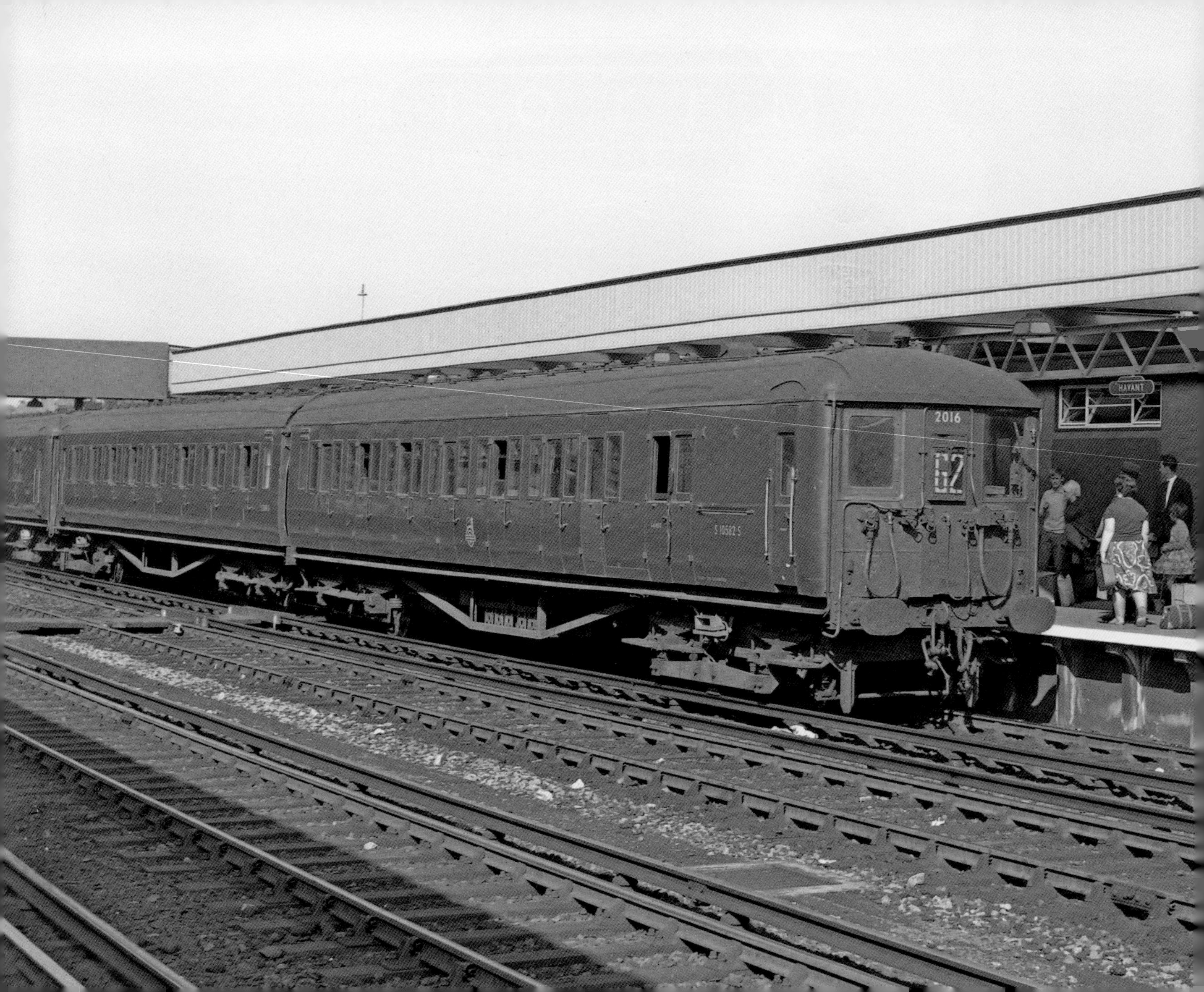
HAVANT
2016
62
S 10582 S

Summer time at Havant with a through Portsmouth - Brighton working entering the platform. From the look of the passengers most would appear to be day trippers or holidaymakers destined for Brighton rather than Waterloo and in consequence this service will be busy to say the least. 2-Bil No 2016 was one of a total of 152 of these units built between 1936 and 1938, four were destroyed or so badly damaged in wartime that they written off. Post war, other accidents and incidents saw one of the two vehicles in the sets damaged beyond repair, including in units 2056, 2069, 2088, 2100 and 2133. All of these were retuned to traffic now attached to all steel trailers built in the same style as the post-war 'tin' Hals. Even with these replacement trailers seating capacity remained as per a standard Bil.(Nos 2069, 2100 and 2133 were the three 2-Bil units coupled to post-war all-steel driving trailers from the mid-1950s.) *(Kichenside 1794 / Transport Treasury)*

The unmistakable outline of a 5-Bel set as used regularly on the 'Brighton Belle'. Introduced in 1934 as the 'Southern Belle', the electric all Pullman service replaced the steam hauled 'Southern Belle' and was renamed the 'Brighton Belle' a year or two later. Over the years it enjoyed regular clientele including Thespians resident in Brighton who would use the train to and from their London engagements. Three units were built, originally as Nos 2051-3, and renumbered 3051-3 in 1937. Operated in conjunction with the Pullman Car Co., The SR, and later BR, were responsible for the underframes, bodies, and electrical equipment plus running the train, whilst Pullman took care of the bodywork and interior fittings. Although associated with the Brighton line a five car set might also be seen on special duty to Portsmouth from time to time whilst there were also regular turns to Eastbourne. In their final years the trains were painted in a lined version of standard BR blue/grey, which did little to enhance their appearance and if anything served to emphasize their age. ('Reversed' grey/blue livery was used on the Blue Pullman and hauled Pullman stock on the ER and LMR [and briefly a few Golden Arrow cars], which looked hideous). There was also the famed incident of the 'kippers', BR taking these delicacies off the breakfast menu which led to howls of protest in the press leading to their reappearance on the menu. The trains ran until April 1972. This pair of images show the regular Brighton Belle train passing Haywards Heath. *(Kichenside 1899 (above) and 1900 (right) / Transport Treasury)*

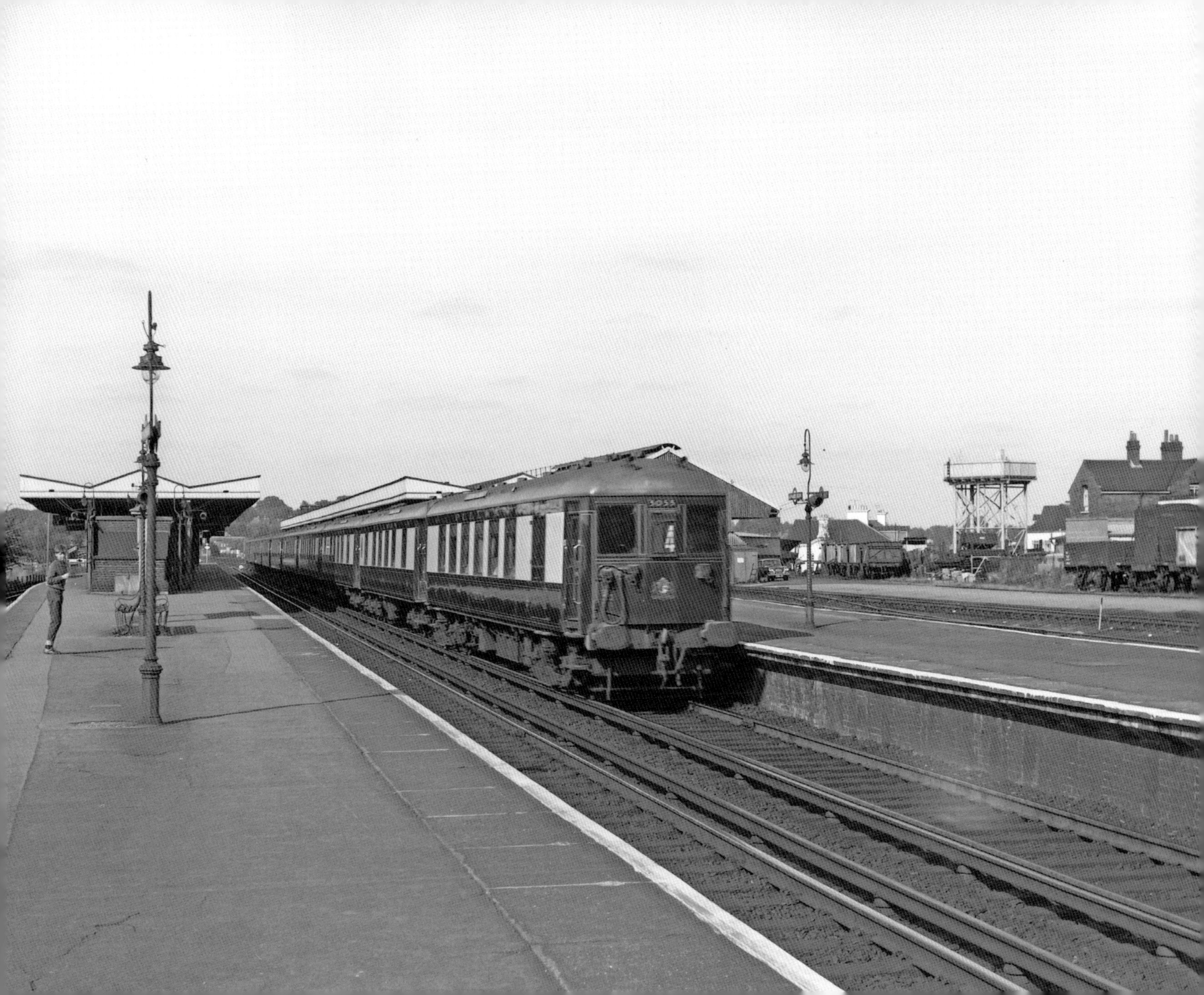

5055
4

Back to Havant where unit 2078 leads a trio of two car sets; 2-Bil / 2-Hal, 2-Bil on what was later to be known as a coastway service to Brighton. Once more it also looks as if the train is going to be busy. Compared with the image on pages 58/59, this time we also see the corridor side of the leading set. *(Kichenside 1710 / Transport Treasury)*

Possibly taken on the same day but not the same train - careful study reveals the first class accommodation on the train seen opposite to be in coach five - this is 2-Bil set No 2058 where first class in the last coach, from its external condition it is probably also fresh from overhaul / repaint. Notice the white dot, referred to on page 33, is not present. *(Kichenside 1706 / Transport Treasury)*

2-Bil No 2034 waits at Haywards Heath with one of the regular services to Seaford, possibly having originated at Horstead Keynes; the driver either wanting to be in the photograph or perhaps unhappy he is being recorded! Aside from services to Seaford, electric trains operated from here to Brighton and London. There was also a shuttle - sometimes extended to Seaford. *(Kichenside 1792 / Transport Treasury)*

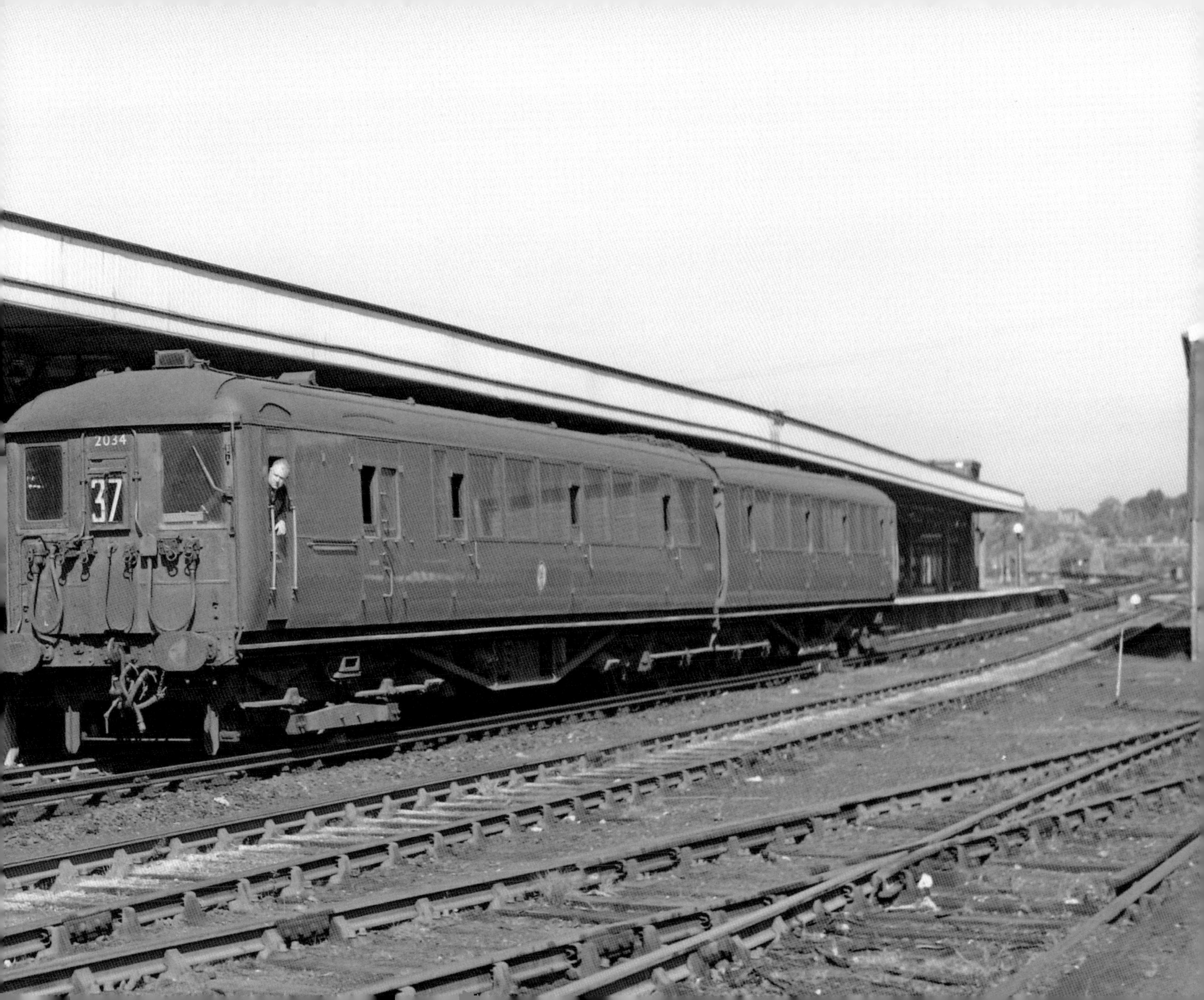
2034
37

Driving motor brake (usually simply referred to as a motor brake second or, in Southern Railway days, a motor third brake), of 2-Hal No 2650. Without any pretence at being cruel, one cannot but notice the hair style of the passenger in the first compartment of the Hal - so much of its time. Construction of these sets was in the traditional wooden body with steel panels and a wood / canvas roof. Although noted for their reasonable riding quality, the seats in third class were changed to a more utilitarian suburban type and the sets came to be regarded as some of the most uncomfortable on the network. Indeed it was debatable if first class accommodation in a Hal was up to the previous third class in a Bil. *(Kichenside 1696 / Transport Treasury)*

S12072S driving trailer from set No 2049. Pick-up shoe gear was only provided on the outer bogies of these 2-Bil sets whilst traditional bogie step boards continue to be fitted. *(Kichenside 1702 / Transport Treasury)*

Believed to be at Three Bridges - the Q1 and numbered pull-push set in the far background are the give-aways. Coach S10782S is from 2-Hal No 2664, in service until 10 January 1970. The chalked / painted lettering on the bogie, 'B-12-9' refers to bogie overhaul. Plates on the end show this vehicle as 'Restriction 4' with a tare weight of 44 tons, the heavy amount attributed to the electric gear / traction motors etc. Notice also the 'tell-tale' on the coach end to indicate if the communication cord had been pulled. *(Kichenside 1700 / Transport Treasury)*

'Modern' 4-Sub No 4715 descending the gradient into Twickenham. The 4-Sub units might be seen on any part of the SR suburban network and often comprised a trailer second (compartment stock), a trailer second saloon, and a motor saloon brake second at either end. There were also variations; all compartment, semi-opens and some had saloon motor brakes and two compartment trailers. In their final years the sets were reformed to consist of just saloon stock in an attempt to deter vandalism which might otherwise take place in an enclosed compartment. White painted steel door handles were also fitted. After their passenger carrying days were over, at least one rake of three sets - 12 cars - was stabled at Eastleigh and used in connection with loco hauled air brake training involving Class 56 diesels. *(Kichenside 1894 / Transport Treasury)*

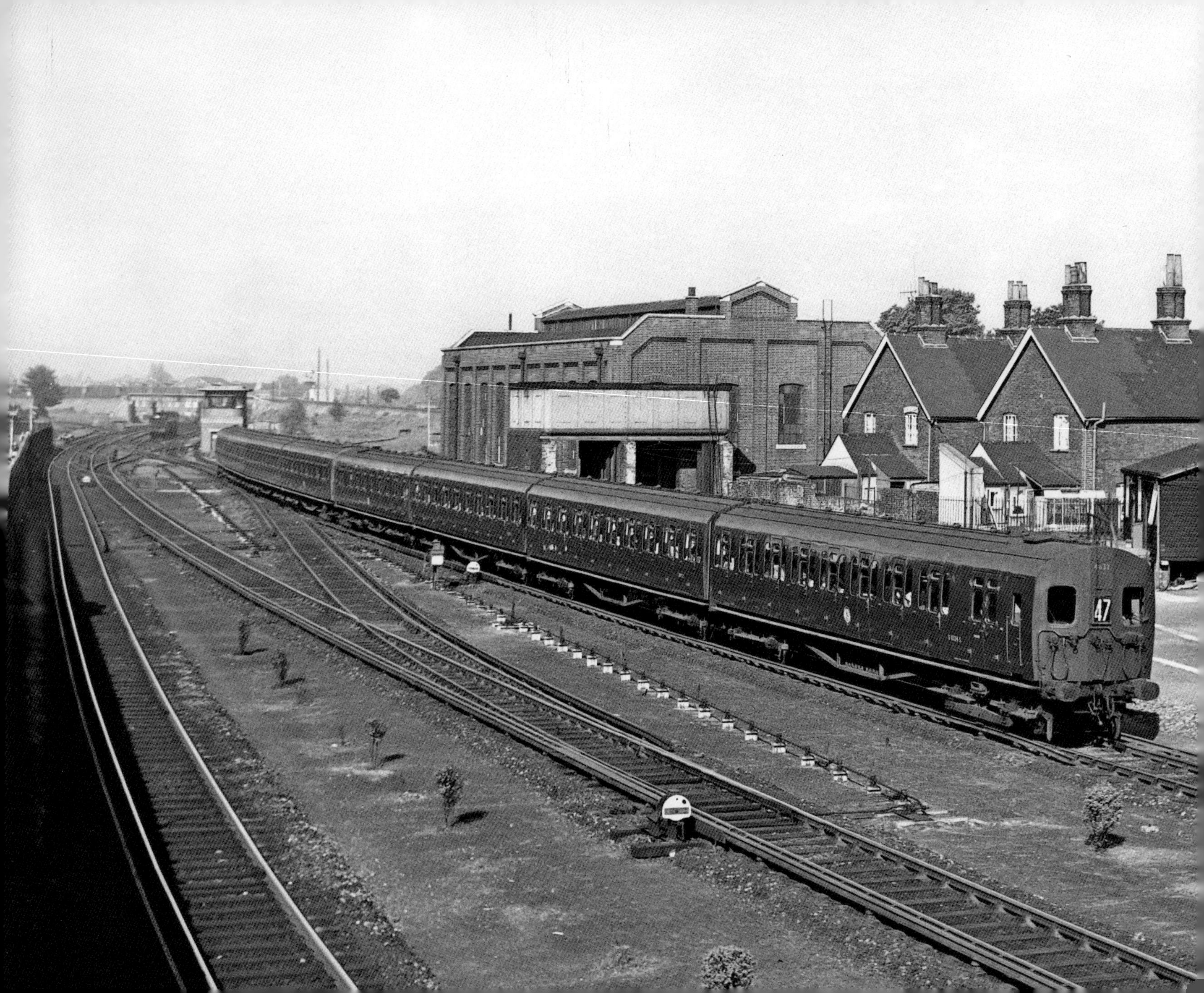
47

The suburban service of the late 1950s and early 1960s; green trains and the BR heraldic roundel with no yellow ends. A pair of four-car units, led by No 4632 on a '47', Shepperton to Waterloo via Richmond service. The train is seen approaching Twickenham. *(Kichenside 19001 / Transport Treasury)*

4-Sub No 4720 forms the rear portion of a train here at Teddington. Although the 3rd rail is in place and indeed has been the some years, in many respects the station retains much in the way of a timeless quality; the station gardens, goods yard with a number of BR road vehicles, pole route, Southern lamp shades and outside seats. Note too the cautionary white line on the edge of platform with most of the concrete lamp standard also painted white up to a set height. *(Kichenside 1896 / Transport Treasury)*

4-Cor No 3153 on the down slow line at Woking heading west with a '72' Waterloo - Aldershot - Farnham working. Alongside a Bournemouth or West of England loco hauled train passes on the down fast. It would be interesting to be able to read the roof boards, possibly Waterloo - Farnham, Cor type units undertaking these services at peak hours. In the background is Woking station and a suburban unit waiting in the sidings. *(Kichenside 1635 / Transport Treasury)*

The new era. 4-Cep unit, No 7116 built for the Kent Coast electrification. Very much based on the BR standard Mk 1, these sets were originally turned out in EMU green and with the roof at the front end protruding slightly over the end of the cab giving a particularly pleasing appearance compared with the stub end of later units. A total of 111 of these 4-car trains were built along with 22 4-Bep units, Nos 7001-22, the latter able to provide catering services. Notice there is no separate driver's door with entry instead through the guard's compartment. *(Kichenside 1903 / Transport Treasury)*

An open motor brake second was provided at either end of the Cep units, the remaining two vehicles being a corridor composite and a corridor second, total seating capacity was 200 second and 24 first; again indicative of how fashion changes had reduced the need for first class accommodation. Only the trailer composite had a toilet at each end. In the trailer second, both were at the end adjoining the motor brake second, as illustrated; one positioned on each side of the carriage, the same arrangement as a loco-hauled corridor or open second. From the solebar this is set No 7110 possibly on a test run soon after delivery in 1958. *(Kichenside 1904 / Transport Treasury)*

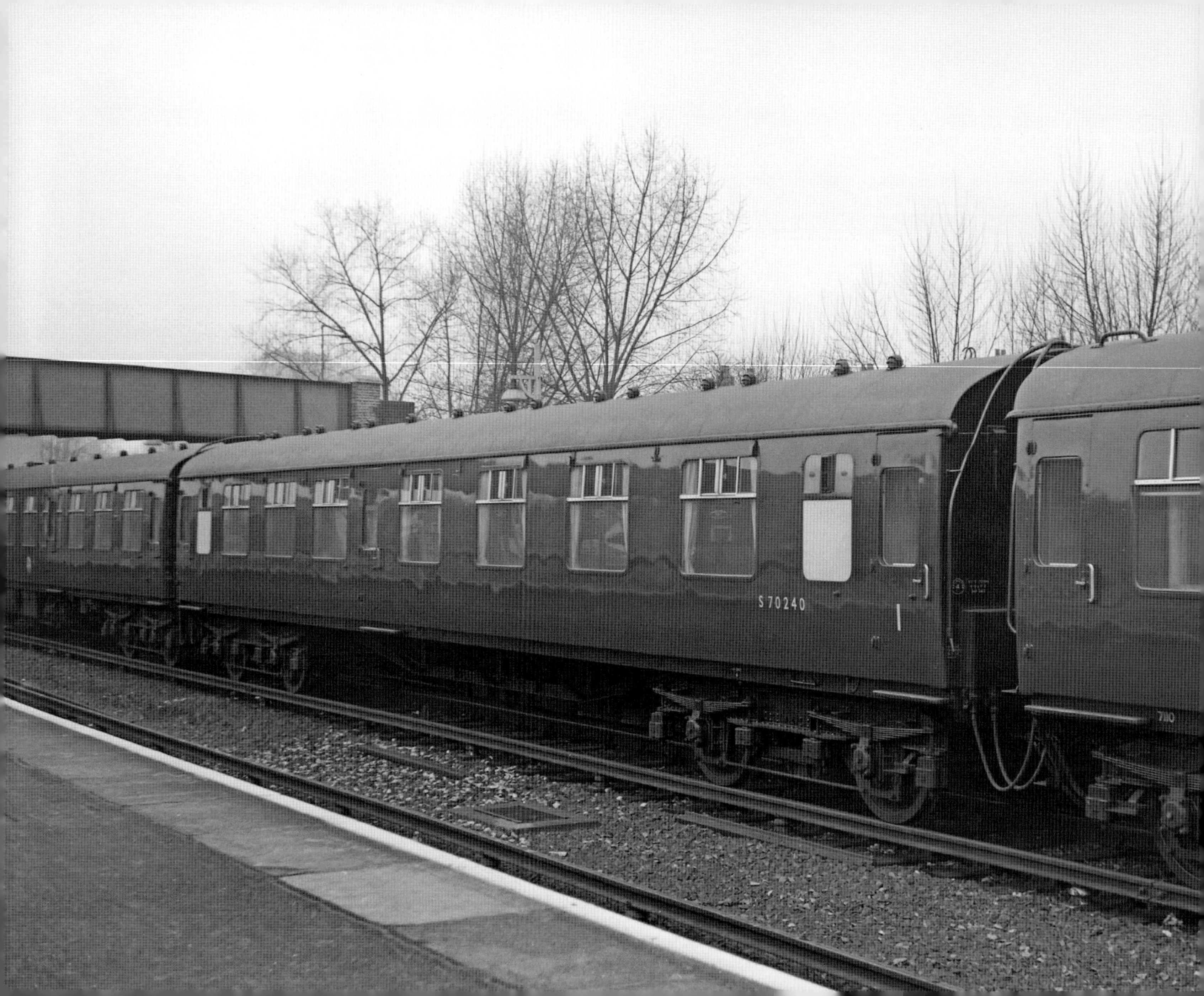
S70240
7110

Two views of the intermediate vehicles in a Cep unit; opposite the corridor composite, and above the corridor second. Thirty years later the Wessex Electric '442' sets would still incorporate some compartments for first class accommodation, folklore having it this was for security purposes should senior staff from the atomic energy establishment at Winfrith in Dorset require privacy. Notice the lack of buffers between vehicles, the unit now having rubbing plates with the coaches semi-permanently coupled. The similarity to Mk1 stock of the period persist with the pipes for filling the lavatory roof tanks at the end of the coach. *(Kichenside 1906 (left) and 1907 (right) / Transport Treasury)*

Buffet vehicle in a 4-Buf unit. This car was originally in set 3073, later 3083. Into service in July 1938 it had a service life of 32 years before being withdrawn on 21 November 1970. The table lamps give a nice touch. Mr Bulleid added his own mark to the interior of some of the buffet sets with a stylish interior intended to woo custom. It did, but perhaps a little too well, some passengers preferring to spend their whole journey in the buffet lingering perhaps over a single cup of coffee and so preventing others from enjoying the facilities. In his next attempt at catering accommodation he perhaps went a bit too far the other way with the windowless restaurant portion of the 'Tavern' cars. *(Kichenside 1714 / Transport Treasury)*

Nine-compartment trailer third from an early 4-Sub built with the intention of being used as a composite: ventilation is either the droplights or the vents but only above three compartments - quickly needed with a compartment full of bodies (might these also have once have been designated for smoking?). Coach S11484S was incorporated in set No 4114, built in May 1946 and in service until 13 May 1972. As before taking a moment to study the expressions and behaviour of the passengers is worthwhile. *(Kichenside 1682 / Transport Treasury)*

BR days and with an 'S' prefix on EPB set No S 5053; the designation EPB standing for 'Electro Pneumatic Brake'. The set is at Dunton Green and again an image with plenty of Southern 'furniture' in view including the barley-twist lamps. Set No 5053 has lost its power jumper. A modification did not take place until the mid-1950s, No 5186 built in 1955 being the first not to be fitted with them from new. *(Kichenside 1628 / Transport Treasury)*

Bibliography / Acknowledgements

The following have been consulted:
The unpublished notes and listings of the late 'Ted' Crawforth.
The SEMG website for details of unit histories and headcodes.
The 'Blood and Custard' website
Ian Allan *'ABC'* booklets - various.
Southern Region Multiple Unit Trains, Gregory D Beecroft and Bryan Rayner. Published by the Southern Electric Group 1979, 1981, 1984.
LSWR Carriages in the Twentieth Century. Gordon Weddel. OPC 2001.
Southern Electric, a new history. Vols 1 and 2. David Brown. Capital Transport 2010.
Transport Treasury would also like to thank Robert Carroll and Dennis Troughton, for their kindness in checking the text.